BHOPAL DISASTER

An Eyewitness Account

Lalit Shastri

Writers' Dream Publications

Copyright © 2020 Lalit Shastri

All rights reserved

No part of this book may be reproduced, or stored in a retrieval system, or transmitted in any form or by any means, electronic, mechanical, photocopying, recording, or otherwise, without express written permission of the publisher. First Edition of this book was released in 1985-86

ISBN: 9798576298600

Writers' Dream Publications
#34 Sagar Garden Homes, Kolar Road, Bhopal 462016
celebrateglobe@gmail.com

CONTENTS

Title Page	1
Copyright	2
A Few Words	5
Introduction	6
Death Strikes	9
Endless Suffering	20
A Total Lie	33
The Culprit	41
Legal Wrangling	45
Justice Delayed	53

A FEW WORDS

Ever since I gained my senses I have been a part of Bhopal. I have grown here, while the city grew around me. Walking down memory lane, I am reminded of the pleasant moments I have shared with this lovely city.

My journey through a short span of time has been quite eventful. I have seen Bhopal prosper - from a sleepy little town into a bubbling State capital and an upcoming industrial centre. Everything was so smooth and peaceful till the intervening night of 2nd & 3rd December 1984, when tons of deadly Methyl isocyanate, stored at the local Union Carbide Pesticide Plant leaked into the air and killed thousands of innocent people.

I saw the devastation caused by the gravest of all industrial disasters right before my eyes and decided to write down my encounters with death and destruction. Through this work I have tried to project the sufferings of the victims before humankind.

LALIT SHASTRI

BHOPAL 15th June, 1986

INTRODUCTION

Bhopal is a sprawling city. It is famous for its lakes and hills. In 1956, it became the capital of the reorganised state of Madhya Pradesh. Over the years the town has witnessed considerable growth on both socio-economic and cultural fronts. During this period its population increased nine-fold and presently (in 1985) the figure stands well above nine hundred thousand.

On a misty winter morning, as the Indian Airlines' Boeing loses altitude, leaving behind the veil of clouds to land at Bhopal airport- the passenger is enchanted by the lush green landscape that comes to his view through the little rectangular window. The channel music stops and the air hostess goes on to announce that in the next few moments the plane would be landing at Bhopal airport. The plane circles the sky twice before coming down to land. It is a treat to look out of the window while the aircraft banks on its starboard wing. On looking out, one gets a clear view of the vast expanse of the lake that magnificently penetrates deep into the heart of the city on the one side, and gently washes its outer limits on the other. On the far side of the lake stands the Palace of the erstwhile Nawaubs of Bhopal and the vast Hamidia Hospital complex, and on the other stretches Shamla hill as if challenging the grand presence of the lake but the beauty of the fresh blue waters remains unsurpassed.

Moving the eyes off the lake one sees the synthesis of the old and the new. Untidy lanes and clustered houses surround a huge Islamic structure with its pair of towering minarets- it is the Tajul Masajid claimed to be the biggest mosque in Asia. Behind Shamla hill, in the south-east direction is new Bhopal. It looks characteristically green with well planned roads and dotted with a modest number of high rise buildings. As the aircraft turns around, the Bharat Heavy Electricals factory and the township come into sight on the eastern side of the main railway line that passes through the city. Moments later, one is flying right over the railway station and a few yards away once can see the Union Carbide Plant. Another minute and a perfect two point landing brings the plane to a halt. Baggage clearance takes a short while and one is soon ushered into a virgin countryside along with the rest of the passengers. A cab is always available for everyone and one soon rushes along a short stretch of level road that curves its way through cultivated fields, before taking a hairpin bend to climb the steep 'Lal Ghati'. As the road levels out, one gets a close glimpse of the city. The town wears a peaceful look. There is a peculiar sign of content on the faces of those one crosses on the way.

The city is characteristically divided by the twin lakes into two halves - the north and the south. New Bhopal - the southern half - boasts of good town planning and is dominated by government establishments. The Secretariat, the Legislative assembly, and all the major schools and colleges are located in this part of the city. Most of the government employees also live here. In comparison, the old city wears a congested look. The railway station, the city bus station, major industries and banks, timber and saw mills, transport offices, cinemas and hotels are in this part of the city. Since most of the commercial establishments are located in old Bhopal it has significant economic importance. This is the main reason for an excessive population pressure here. There is a constant influx of immigrants who come to the state capital, hunting for jobs. A concentration of mobile population, in the old city, has given rise to temporary shelters that have gradually taken the shape of slums, particularly near the railway station, bus station and factories.

Overlooking the population pressure and the growing needs of the town, the state government allotted a site to Union Carbide India limited to establish a potentially hazardous pesticide plant at a stone's throw, barely eight hundred yards, to be precise, from the railway station.

Shanties soon sprung up near the Carbide plant and during the last ten years slum towns like Jai Prakash Colony, Kenchi Chola, Bapna Colony, and Bidi Walon Ki Basti, grew around the plant. Unfortunately, Jai Prakash Colony stretches right across the main gates of the deadly plant. The population in these colonies soon exceeded fifteen thousand and Arjun Singh, former Chief Minister of Madhya Pradesh-in one of his irrational moves, granted lease or call it settlement rights to each of the slum dwellers. When disaster struck on the midnight of 2nd December 1984, it left most of these inhabitants dead.

DEATH STRIKES

Today Bhopal looks visibly shaken. It has been through a calamity that has left thousands of its citizens dead and more than a hundred thousand critically affected. Entire families have been wiped out, countless children have become orphans overnight, and many have survived only to lament the loss of their loved ones. Livestock has perished in great numbers. All as a consequence of the leakage of deadly Methyl Isocyanate gas from the Union Carbide plant on the night of 2nd and the early hours of 3rd December, 1984. According to specialists the world over, never before has such a toxic substance been released in such huge quantities - certainly not with such devastating effects.

That deadly night, Bhopal went through the worst industrial disaster ever. The modern world has witnessed many industrial mishaps in the past. On September 21, 1921 an explosion at the Base Chemical Plant at Oppau in West Germany killed 561 people. On November 19, 1984 at least 452 people died in Mexico when 80,000 barrels of natural gas exploded at a state owned Pemex Facility. These accidents pale into insignificance when compared with what occurred at Bhopal. During those few hours, in the cold stillness of the night, Bhopal had turned into a giant gas chamber.

On that dreadful night, death ruled the city; there raged an unequal battle between life and death. Thousands perished within a matter of hours. The dead lay strewn in the streets and alleys - they had collapsed while running for their lives, others died in their sleep unaware of the lethal cloud that silently crept upon them. Entire families lay dead. Death struck down not one, not a few, but countless innocent and unsuspecting men, women and children. The memory of this disaster shall haunt us forever.

In Bhopal, winter is at its peak in December. It rains very rarely and the skies remain remarkably clear. During the day the warm rays of the sun offer some respite from the cold but with nightfall the cold is bone chilling - those early hours of 3rd December were no exception. It had been unusually cold that day. People had either retired after a hard days' work or else were about to do so. The settlers around the Carbide plant had no inkling of the deadly events that lay ahead of them.

It was the residents of Jai Prakash Colony (a settlement of about 800 houses) and the adjoining shanty towns that were the first victims of the escaping gas. Around midnight, some residents woke up all of a sudden feeling suffocated and seized by a fit of coughing. As they opened their eyes they felt an intense burning sensation, it was as if ground hot-pepper had been sprinkled in their eyes setting them on fire. Unable to locate the cause of this sudden discomfort, they ran out of their houses into the open to catch some fresh air only to see a strange cloudy mass suspended above them, but there was no respite from the invisible flames that were searing their lungs and eyes. Within a matter of moments, anyone who had not succumbed in his sleep was awake and running in all directions, wailing and shouting.

Instant panic spread in the low-lying areas adjoining the Carbide Plant. People rushed out of their doors. The crowd soon swelled onto the streets and the exodus led to a virtual stampede. People ran for their lives without knowing what they were running from. They were gripped with fear and unable to fathom the cause of their misery. Every breath they drew enhanced their pain. In those moments, people lost faith in the air they had always breathed. That very air had suddenly turned into something fearsome and was bent upon destroying them. The terror was so great and the desire to flee so acute, that some people even left their children as they fled. Families running for safety could not hold on together for long. They ran wherever their feet took them. Some managed to escape to safer places but many died on the streets.

The railway station which is situated at a mere stone's throw from the Carbide plant had its own tale to tell. Many railway employees working the night shift, porters hoping to make an extra buck on the early morning trains, and beggars to whom the hallways of the railway station provide some shelter from the chill outside, were all found dead the next morning. The death toll would have been much greater had it not been for the Railway superintendent and a senior booking clerk. Bhopal is a major railway junction; it lies on an important trunk route connecting the north and the south as well as the East and the West, and many trains carrying hundreds of passengers roll into Bhopal station every day. The two railway employees had the presence of mind to realize what would happen if trains scheduled to arrive in the early hours of the 3rd morning were allowed to come in. Both stuck to their posts and breathed their last sending messages to adjoining stations to stop all incoming trains. They have left behind a saga of selflessness, exemplary service, and dedication to mankind.

At J.K. Straw Products, a paper factory situated close to the Carbide plant, a hundred and seventy six employees worked on the night shift ignorant of the imminent danger. The lethal gas soon pervaded the plant and the workers began to experience its effects. The management of this plant however, acted quickly and sent an S.O.S. to the local Army Center. The army responded promptly and due to this timely help most of the workers were safely evacuated to nearby hospitals. However, not everyone could be saved and many succumbed to the fatal effects of MIC.

The leak of poisonous gas from the Union Carbide plant could have been sealed promptly without much damage but for gross negligence on the part of the staff on duty that night. The leak was noticed by the plant operators at 11.30 pm on 2nd December but was ignored! They thought it to be a mere 'routine leak". The Production Assistant Shakil Ibrahim Qureshi was informed of it and the matter simply rested there - no measures were taken. From then onwards, until 1 am of 3rd December (0100 hours), when untreated Methyl isocyanate vapours were seen escaping through the 33 metre high flare tower, there was a complete breakdown of all norms of safety procedures. The public warning siren was sounded at midnight but was switched off almost as soon as it had gone on. Subsequently, the muted siren - meant for the plant employees only, was sounded. The public siren came on again only at 1 am. The warning came too late - the escaping gas had already taken its toll. By this time the gas was escaping in huge quantities, hundreds had already died in their sleep, many more had fled - some had even reached the hospitals for medical help!

Admittedly, this was an unprecedented situation. However at each step we find that there was a complete mismanagement of affairs. At the time of such a grave crisis, it was not just the Union Carbide but also the city administration that miserably failed in the task of crisis management. Instead of gathering their wits together to cope with this grave disaster, they acted in a manner that is most unbecoming of the responsibilities reposed in them. The administration was informed of the gas leak at 1 am. Moti Singh, the head of the district administration, who should have immediately galvanized the official machinery into action fled from the scene to a safer place where he later established the control room. The list of public officials who ran off to save their own lives is a long one and it is common knowledge that even the then Chief Minister Arjun Singh was on the run; he took refuge in the government owned luxury guest house at Kerwan Dam, a fair distance from the Carbide Plant. Arjun Singh later categorically refuted this allegation.[1] However, there is some strength in the claim that Arjun Singh had fled as he was neither seen nor heard conducting relief operations until 3rd morning. Most of his Cabinet colleagues and high officials in the government were no exception either. What ensued was a total breakdown of the state government machinery. Those fleeing from the gas affected area received negligible help and guidance from the authorities. All were left to manage on their own, there were no transport facilities available to carry people to safer areas, and nothing was announced over the public address system until about six o'clock in the morning, when police vans with mounted loudspeakers were pressed into action. But alas, all they had to blare out after thousands lay dead was:

"Something had gone wrong somewhere. Everything is normal now Citizens are requested to return to their homes."

Thus, a majority of the people affected by the gas did not know the cause of the disaster; they did not know the direction in which to run in order to escape from the poisonous gas. The only help they received was from the Army and a few brave citizens who volunteered to commence relief activities. The Army carried on uninterrupted rescue operations and its personnel risked their lives to carry hundreds of gas victims to the military and government hospitals. While doing so, two Army drivers succumbed to prolonged exposure to MIC fumes. As dawn broke, private vehicles also started carrying people away from the affected areas to the hospitals. It was only around this time that the official machinery got into gear and commenced relief operations in true earnest.

One wonders how such a terrible accident occurred and that too at a supposedly modern multinational plant that boasted of its U.S. connection. The probable factors that ultimately led to such grave consequences can be traced to unforgivable human lapses and an unacceptable sequence of system failures. On the night of 2nd December there was an abnormal build-up of pressure in tank 610, one of the three underground tanks containing MIC at the Carbide Plant. Shortly after midnight, the pressure reached the ceiling value of 40 pounds per square inch (PSI) and the rupture disc that is supposed to burst at this pressure gave way. Subsequently, a release valve on the pipeline going out of the MIC tank - also popped. The next check point was the release valve that is meant to conduct high pressure gas into a vent scrubber via a 70 feet long pipeline. The scrubber is a device containing packed rings of caustic soda and the gas under high pressure is supposed to percolate through these rings. The gas reacts with caustic soda and is detoxified in the process. That night, since the pressure of the gas was very high it passed through the scrubber too fast for this to happen. Besides, the caustic pump was also not operational and hence the vent gas scrubber was not charged once it was exhausted. Thus the unneutralised gas surged onwards to the 33 metre high flare tower. This tower performs the function of burning toxic gases before they are finally emitted into the atmosphere. However, it has come to light that on the night of 2nd December the flare tower was under maintenance and hence did not perform its all important function.

The chain of safety devices also included a cordon of water sprays that are capable of shooting water jets 12 to 15 metres into the air causing the water to fall on the escaping gas and neutralizing it. Strangely enough, these jets were designed to shoot water only to a height of 15 meters whereas the flare tower through which the gas was supposed to emit was 33 meters in height! Thus, the water jets were not capable of neutralizing gas released into the atmosphere through the flare tower.

The Union carbide manual on MIC states that "pressure in the tank will rise rapidly if MIC is contaminated and in order to contain violent reaction bulk systems must be maintained at low temperature." According to the manual, the MIC storage tanks are connected to a thirty ton refrigeration system to circulate the liquid MIC keeping it "preferably at 0 degrees Celsius. The chiller in the plant was so adjusted that it allowed the temperature to rise as high as 20 degrees Celsius before an alarm was raised. This was in complete violation of Carbide's 1976 technical manual that warns specifically that if MIC is kept at that temperature a contaminant can spur a runaway reaction. As if this was not enough, on the night of the gas leak, the refrigeration unit was not even functioning.

Following the gas leak, Union Carbide has reiterated on a number of occasions that a possible ingress of water or some other substance into the tank resulted in the excessive pressure build up. However, they have not offered any explanation as to why none of the safety devices were functioning in the manner they should have. As far as the operating staff is concerned, we know for certain that they did nothing to check the cause nor took any corrective steps after they noticed signs of an increase in pressure.

The possibility of contamination and subsequent systems failure indicate that the plant was being operated improperly and also that there were serious defects in its design. The Union Carbide Corporation has repeatedly said that long before the leak of MIC, they had brought to the notice of their Indian counterparts that the safety measures were not being properly followed at the Bhopal plant. This claim of Union Carbide Corporation does not in any way absolve it of charges of criminal behaviour because it did not take any steps to stop the plant from continuing production in spite of being fully cognizant of the unacceptable state of safety system at the Bhopal plant. The Union Carbide Corporation had every reason to know the unlimited hazards the residents of Bhopal faced from a possible leak of MIC.

The role of Union Carbide, ever since it set its foot in Bhopal, has been nothing short of criminal. In the early '70s, it developed a fully computerised four-stage alarm system which is operational in its West Virginia plant. However when it set up its plant in Bhopal, it actually transferred a second-hand and obsolete MIC plant from Danbury USA that had a substandard safety system consisting of manual controls with only one back-up alarm system. The pressure gauge meant to measure pressure in the MIC tanks was also monitored manually whereas; in the West Virginia plant it is linked to an automatic warning system and the control room. Many valves and vent lines in the Bhopal plant were too old and worn out. Most parts that should have been replaced every six months had not been changed for over two years. All this in an effort to cut operating costs. For the same reason, Union Carbide adopted a manufacturing process that involved storing large quantities of a lethal chemical; whereas plants abroad -particularly the one in West Virginia, employ a process that does not require storing MIC in huge quantities. MIC produced in these plants is immediately converted into end products that can be stored safely.

The Union Carbide Corporation is guilty of gross disregard for human life as it did not take any steps to warn the authorities and the citizens of Bhopal about the extremely lethal nature of MIC and its hazards. It is a serious crime on its part not to have directed the local police and the civic administration to conduct emergency evacuation drills at frequent intervals near its plant. The company should have informed them about the dangers the residents of Bhopal were facing from a possible leak of MIC so that they could have taken appropriate measures to combat any crisis arising out of its leak.

The company also kept the fire department totally in the dark about the toxic nature of the gas. The fire department was unaware that spraying water on MIC would neutralize it. It remains a mystery why on the night of the gas leak, the local management did not approach the fire department for its help. Fire tankers should have been mobilised to spray water in the gas affected area to substantially neutralize the escaping gas.

Another offence committed by Union Carbide was that it did not communicate to the health authorities of Bhopal the harmful effects of MIC nor did it inform them about possible antidotes in case they had to treat someone exposed to the toxic gas. Indeed, when the victims of MIC poisoning arrived at the hospital, the doctors were unaware of the best form of treatment - the uncertainty and confusion prevailed for more than two months and compounded the tragedy and human suffering.

Finally, one is confronted with the false stand taken by the Carbide management of Bhopal following the gas leak. The Works Manager J. Mukund told reporters on 3rd morning, that the leak had been sealed within minutes, whereas, the entire tank containing 15000 gallons of MIC had emptied into the air.

The Union Carbide Corporation is criminally culpable of mass homicide because, although it was aware of the hazards of MIC and was taking every possible safety precaution at its plant in West Virginia, it did not take adequate steps for the safety of the people of Bhopal and kept them unaware of the potential threat to their lives.

Even the Union of India and the Government of Madhya Pradesh cannot escape charges of wrongly permitting Union Carbide to use such a deadly and lethal gas within city limits. Both are guilty of criminal negligence on many counts namely - a) not caring to inform themselves about the hazardous nature of MIC and allowing Union Carbide to store it in huge quantities[2] within city limits, b) allowing a factory to carry on production while most of its safety mechanisms were out of order and even otherwise inadequate, c) not taking any effective steps in spite of the deaths of at least three workers and repeated reminders from various sources regarding the dangers faced by the citizens of Bhopal because of the Union Carbide Pesticide Plant, and d) permitting the growth of slums right across the Union Carbide plant and subsequently granting permanent lease to the slum dwellers.

The leak ultimately caused the deaths of thousands of inhabitants of Bhopal. Every compassionate person has been shaken by this industrial disaster. We have long been aware and concerned about the dangers that accompany the use of atomic and nuclear energy but the deaths at Bhopal remind us of the horrors that can stem from improper use of chemicals. One has remained complacent far too long about the dangers of the large scale use of lethal chemicals. Bhopal has demonstrated that we must change our attitude and go for extremely stringent safeguards so that there may never be another Bhopal in the future.

Man has forever been at the mercy of nature. All his efforts in his walk towards progress have been geared at controlling the natural forces thereby making the world more liveable. Our scientists are constantly engaged in the task of preventing natural calamities. They are studying the laws of nature and trying to understand the temper of natural forces continuously at work. Their efforts go a long way in predicting earth quakes, hurricanes, floods etc. Attempts are perpetually being made on a global scale to save human life by developing early warning systems. In this backdrop, an industrial disaster like the one that occurred in Bhopal is a sheer paradox. In Bhopal, a giant multinational corporation built a potentially dangerous chemical plant, with no consideration for the lives of the innocent citizens of Bhopal who ultimately had to pay a very heavy price for the unpardonable crime on the part of the torch bearers of progress.

Today questions loom ahead of us - Development, but at what cost? Do we have to walk over the dead for the sake of progress? People sum up industrial calamities as the price for progress. They justify the presence of Union Carbide in Bhopal on the grounds that it was engaged in producing insecticides that are essential for producing bountiful crops to feed the teeming millions of India! Such an argument is misleading because it distracts from the real issue. One is not against producing insecticides provided this can be done while assuring the safety of people it is intended to help. It is the inhuman callousness of big multinationals that operate unsafe plants in order to cut operating costs that needs to be brought into focus. This is particularly true of such plants operating in the developing countries. Union Carbide for instance, delivered for Bhopal an outdated plant design with manual safety controls although similar plants in developed countries use fully computerised safety systems.

Furthermore, some of its critical plant equipment was in poor condition and no efforts were made to replace it.

There are many reasons why the multinationals find the third world a fertile ground to dump their dangerous production units and walk away, unscathed, killing thousands of people. In a developing country like India, the need to attain self sufficiency in a crucial area like food production is immense. The urge to progress and catch up with the developed world is the need of the hour. But, since the country lacks advanced technological know-how and trained manpower, any offer by leading multinationals to set up plants in India is accepted without close scrutiny. The multinationals exploit the situation by adopting manufacturing processes and plant designs that reduce costs but at the expense of the safety and health of hapless workers and local residents. Even when mishaps occur in such plants due to inadequate safeguards, faulty design and engineering, the people blame their fate and sum up the gravest of tragedies as "God's wish". They fail to understand the crux of the problem.

On studying the modus operandi of the multinationals we find that they use their immense monetary resources very effectively to influence decisions at the local level. They resort to pay-offs to politicians and bureaucrats in order to bypass existing regulations and to obtain improper certificates of inspection. The consequences are for all to see. As late as today, many claim that the gas leak at the Carbide plant was accidental and it was due to mishandling on the part of the plant operators whereas facts reveal a lot more. The period following the Union Carbide disaster has undoubtedly demonstrated the fact that vested interests leave no stone unturned to make sure that the bad press following the worst of catastrophes wanes out before radiating very far. Soon all is forgotten and the culprits who fail to guarantee the safety of the citizens reign supreme.

A month after the tragedy the state government declared that 1422 persons and approximately 2000 cattle and goats had died after inhaling MIC. News reports claimed that 2500 persons perished. The real number is many times more.

Even today gas victims are dying in the hospitals. Many have been blinded and yet many more are gradually losing their eyesight. Many pregnant women have miscarried. Other harmful effects remain to be investigated. The disaster has already taken place but the aftermath is far from over.

ENDLESS SUFFERING

3rd December 1984, 8 o'clock in the morning. It was time for the news bulletin on All India Radio. I could hear the headlines - there had been an explosion around midnight at a gas plant in the old section of Bhopal... the damage was still being ascertained. I was baffled by the news. All at once many questions sprung to my mind: Where exactly was this gas plant? Was the gas harmful (it must be I thought, why else would it be announced on the national news as the top story)? What areas had been affected? Strangely enough, neither the location nor the magnitude of the explosion was reported in the news and I also remember that there were no instructions about evacuating the people from the affected areas.

That morning I was to take the Civil Services Examination being held in the old city. As there was no news about the examination being called off, I made my way to the examination centre assuming that the explosion was not very serious. I was least prepared for the sight that met my eyes as I drove past the Jai Prakash hospital (15 Km from the site of the disaster). To my bewilderment I saw the hospital premises and the footpath across it, flooded with people. I was shocked to see the number of people who were lying on the road - some appeared dead while others seemed completely exhausted and unable to breathe. Most seemed to have vomited and froth appeared around their mouths. I noticed that even those who appeared to be alive had their eyes shut.

As I moved on, I was surprised to find people fleeing in utter panic. They were rushing away in cars, tempos, auto-rickshaws, mini-buses and lorries. They had even climbed on bus tops and crammed inside the boots of the cars. A few of them had managed to find just enough room to hang dangerously on to these fast moving vehicles. These scenes reoccurred throughout the route. Finally, after making my way through this wild exodus, I reached the examination center, totally unnerved, but not too sure about the reason for the panic. (I knew instinctively that this had to be related to the news I had heard in the morning).

As I wrote my examination, I was suddenly gripped with a severe attack of breathlessness and a few moments later my eyes started burning. It was then that I picked up from bits of conversation between the invigilators that some gas storage tank containing a dangerous toxic gas had burst at the Union Carbide Plant and a possibly fatal gas had escaped into the atmosphere. Halfway through the examination, my symptoms became more acute and before I could finish answering all the questions, I was forced to rush to Hamidia Hospital (the main hospital in Bhopal).

The sight I witnessed at Hamidia Hospital was similar to that I had seen in the morning but the panic, despair and agony visible was on a much greater scale. I was shocked to see the number of people who were waiting for medical attention. Thousands of them lay scattered on the floors, passages, roads and pathways and many more continued to stream in. The hospital complex was littered with dead bodies and my mind became numb as I walked through death and pain. I had to struggle my way through the stench and stink of the overcrowded passages to find a doctor. There was no space left for me to walk as victims with vomit stained bodies lay all around me. I soon realized that the doctors were few and they were overwhelmed by the rush of victims who were in much greater need of attention, many of whom were dying within minutes of their arrival at the hospital. The pain in my eyes -though by no means minor, became unimportant before the suffering I saw around me and I decided to leave the hospital without receiving any medical help. The same evening when I went to the Jai Prakash hospital for treatment,[3] I stopped for a short while to talk to Hiralal of Bapna Colony - one of the badly affected localities. He had fled from the scene of the disaster the previous night with a bare minimum of clothes on him. His eyes were red and swollen and he was finding it extremely difficult to keep them open. Panting for breath while talking, he narrated "it was 1 o'clock at night, I had gone to a wedding and while returning home, I was suddenly gripped with an acute sensation in the eyes. The sky was overcast with thick fumes and I ran towards my house. Reaching there, I discovered to my horror that the doors of my house were wide open and both my wife and my five year old daughter were missing. Next moment, I traced back my footsteps as I felt suffocated. Although I was exhausted and breathless, I ran non-stop and covered six miles to the T.T. Nagar area where I fell down on the pavement. When I regained my senses it was morning, I got to my feet and finally reached this hospital. All is finished, my folks... I just don't know where they are?" Suddenly he was quiet, motionless and then breaking the silence he spoke in despair, "Bhai Sahab, Union Carbide has killed thousands of us."

Hiralal's tale of woe had disturbed me and before he finished it I realized that I was surrounded by dazed and fear stricken faces. Each of them had the same story to tell. I took their leave and left for home with a heavy heart.

Next day when I visited the Hamidia hospital, I was able to talk to a few of the doctors. They told me that the first stream of patients started arriving around 1 am and within the next two hours more than two thousand people had already landed at the hospital for medical assistance. On further enquiry, members of the hospital staff disclosed that the superintendent was immediately informed of the emergency. However the superintendent whose residence is just a five minutes' walk from the hospital came after two hours. Those who arrived first, found to their utter dismay that the doctors on duty at the casualty ward had panicked at the sudden influx of patients and all through the night they could not evolve a treatment strategy. Finally after listening to so many people, it became evident that the hospital authorities had never been prepared for such an eventuality. It was no surprise therefore that when faced with a crisis of such a huge magnitude the health administration machinery had cracked down completely.

As victims flooded in, wave after wave, the doctors from Hamidia Hospital contacted the Chief Medical Officer of Union Carbide to enquire about the nature of the gas. He told them that the gas was not poisonous and the patients should be advised to cover their eyes with wet towels. Without divulging the name of the gas, he said that the effect would not last very long. The doctors found the Medical Officer's statement rather strange because the patients were dying within minutes of their arrival. The situation was getting out of hand and in desperation the doctors decided to get in touch with some employees of Union Carbide who were their personal acquaintances. They were informed that two poisonous gases -Phosgene and Methyl isocynate, were stored at the plant. The doctors were already aware that about a year ago three plant workers had died due to a leakage of phosgene in the plant. Hence, they assumed that phosgene was responsible for all the havoc that night. It was later that morning they came to know, according to the Carbide Works Manager that a tank containing MIC had developed a leak that had been "sealed within minutes". The doctors at the hospital were now at a complete loss. They did not have the slightest bit of information on MIC -they did not know how toxic it was and neither did they know what treatment should be given to those exposed to it. Such ignorance was indeed surprising for at least the city health authorities should have been given complete information by the Union Carbide about the gases stored at their plant. Finally left with no option, the doctors and the Superintendent queried each other about the course of treatment. It soon became evident that no one knew what line of treatment to adopt. This added to the panic and anxiety of the patients whose suffering was increasing every minute. Such was the confusion that prevailed during those early hours and in the days that followed.

While the doctors were struggling to determine a treatment strategy, the victims awaiting medical help lay blinded and gasping for breath. The situation soon became uncontrollable as busloads of victims kept pouring into the hospital. The condition of the patients was such that they were dying within minutes, even before they could be attended to. They died of suffocation by spasm of the bronchial tubes that carry air to the lungs. Many had already died in this manner while they slept unto the early hours of the 3rd. The collection of body fluids in the lungs due to the highly irritating gas caused death by drowning. By day break, all the three to four hundred odd doctors of Bhopal were attending to the gas victims and another two hundred and fifty gradually arrived from different parts of the state and country. The students of the local colleges and various voluntary action groups also stepped in to provide relief. The hospitals very soon ran out of oxygen cylinders that were immediately supplied by the Bharat Heavy Electricals Limited in Bhopal and more were flown in from Delhi on a special Air Force plane.

Relief operations were also in full swing but things showed little signs of improvement. In the Ophthalmology department the doctors were overwhelmed by the ever increasing number of patients with symptoms they were not too certain how to treat. Confusion reigned in the hallways all morning as thousands of people lay on the floors writhing in pain with red and swollen eyes. The situation was particularly beyond control in the absence of the head of the department who did not come to the hospital until noon.

The eye specialists, who attended to the victims, found on preliminary examination that the gas had severely affected their eyes. The effects varied from burning sensation in the eyes, watering of eyes, photophobia, and pain in the eyes to defective vision.

On the basis of the initial investigations patients were administered medication to relieve them of their symptoms, however, a thorough check-up later revealed that the problems were more serious than one could imagine. The doctors diagnosed that the patients were suffering from the following problems:

(a) Conjunctival Hyperaemia, (b) Celiary Congestions, (c) Superficial Keratitis [1] [4] (d) Corneal Ulcer (e) Iridocyclitis.

In the last case, the most serious of them all, the gas had penetrated deep inside the eyes, affecting the lens and causing haemorrhage of the retina. I was told by the ophthalmologists that the immediate damage caused by the gas was the destruction of the cells of the cornea or the transparent covering of the eyes. According to them, on coming in contact with the moisture in the eyes, the gas reacted to form hydrocyanic acid which in turn destroyed the cells. Thus the cornea was damaged, in some cases this could lead to permanent blindness.

The effect of the gas on the eyes depended on various factors, mainly, the duration of exposure, concentration of the gas in the atmosphere, distance of the affected person from the site of the disaster and what the affected person was doing at the time of the exposure, whether one was asleep or awake out in the open.

The doctors told me that it was only a week after the leak that experts were able to throw some light on the best line of treatment to follow. As a result till then, the treatment plan for the gas victims remained haphazard as graduate medical students struggled with symptoms that were beyond the comprehension of specialists. The medical profession was left with no choice but to try out various remedies. The gas victims became an experimental field for the doctors. Many of the victims, who were treated with steroids, developed a severe eye condition that may lead to permanent blindness because the steroids reacted negatively causing ulcers in the eye. I saw the doctors administer simple distilled water in the eyes of victims. Health officials have ignored one of the standard definitions of blindness, recognised by the World Health Organisation that states, "Blindness occurs when visual acuity is of less than 3/60 or visual field is less than 30 degrees" and stated that none of those who have been affected by the gas will suffer from total blindness. However the senior ophthalmologists of Hamidia hospital have long concluded that many victims have already gone blind and many more are on the path to blindness.

With the passage of each day, specialists at the hospitals repeatedly impressed upon the fact that the gas that had leaked was Methyl isocyanate and not Cyanide and that it was less toxic than the latter. They even refused to recognize the possibility that MIC could convert into Cyanide inside the human body and adversely affect the optic nerve, either directly or indirectly. In all circumstances it would have been advisable to administer B-complex prophylactically to all patients whose eyes were affected to prevent damage to the optic nerve. However, for reasons yet unknown this was not done and the failure on the part of the responsible doctors to take such a simple precaution will probably cost a lot of people their eyesight in the long run.

Those patients who came in with affected lungs and had difficulty in breathing were also given symptomatic treatment initially. Doctors noted that patients who had been subject to prolonged exposure to MIC, suffered from acute breathlessness and pain in the chest. The highly toxic gas they had inhaled acted as an irritant constricting the airways and destroying the mucosal membranes of the bronchi and the airways. Due to the destruction of the walls there was collection of fluid in the lungs and this phenomenon could be attributed to the increased permeability of arterioles and capillaries. During my investigations, I learnt that x-rays were not taken till three days after the victims started arriving in the hospitals. Radiology reports confirmed that the effect of the gas varied from increased broncho vascular markings to pulmonary edema [5] in the more serious ones. The initial feedback received by the doctors confirmed that MIC belonged to a family of toxins that had no immediate antidote. Therefore, they concluded that the best way to reduce its effects on the lungs was to administer patients with air rich in oxygen and mitigate their pain by prescribing painkillers and sedatives. But due to an acute shortage of oxygen the doctors were unable to administer oxygen to all the patients. I observed that the doctors also gave Atropine injections (Atropine is a dilating agent) to the victims in an effort to combat their breathlessness. Unfortunately, the victims were not dying due to the constriction of the airways but were drowning as their lungs were being filled with body fluids. In order to save the lives of the victims it was essential to reduce the toxic effect of the gas by promptly administering a proper antidote but the doctors were helpless in this matter.

For a considerable period of time after the gas leak, there prevailed a bizarre situation that arose from the stand taken by the State Health authorities and the Union Carbide concerning the effects of the gas and the line of treatment. Both jointly insisted that MIC could never cause Cyanide poisoning and went a step further to reiterate that MIC poisoning has no long-term effects. They maintained this stand even though autopsy reports showed traces of Cyanide present in the blood. The naked eye findings on 3rd and 4th December also revealed signs of Cyanide poisoning, red and pink colour on the body and froth in the mouth. In the first week of December (within a week of the disaster), the State Health administration was informed about the analysis and toxicologists' tests that showed "positive for Cyanides and Amines". Max Daunderer, a well known West German scientist, who arrived in Bhopal soon after the incident, conducted his own tests and announced that Cyanide poisoning could not be ruled out. He advised that Sodium Thiosulphate, considered to be a standard antidote for Cyanide poisoning, be given.

A few days later, the Union Carbide Corporation also sent a message to Bhopal suggesting a line of treatment that included Sodium Thiosulphate. They too did not rule out the possibility of Cyanide poisoning. In mid December, the State Health authorities received another message, a copy of a telex sent to the U.S. Embassy in Delhi from the Centre for Disease Control in Atlanta. This too, like the previous message dealt at length with various problems relating to Cyanide poisoning.

The Madhya Pradesh government finally decided to administer

Sodium Thiosulphate a week after the disaster, but strangely enough stopped it abruptly on 12th December. For reasons unknown, a 'confidential note' was circulated among the doctors banning the use of Sodium Thiosulphate. On 14th December Union Carbide India also issued a statement at a press conference in Bhopal which further hampered the course of treatment.

The statement read: "Methyl isocyanate is not Cyanide. They must be differentiated. Isocyanates are molecules containing the radical NCO, whereas Cyanides contain the CN radical MIC naturally degrades in the environment by reacting readily with water to become harmless substances, while Cyanides do not react with water. These two substances have an entirely different effect on human health."

Contrary to the above information, the guidelines titled, 'Occupational Health Guidelines for MIC; released by the Occupational Safety and Health Administration of the U.S. Department of Labour, in September 1978, state that on reaction, MIC may release hazardous decomposition products, for example, toxic gases and vapour (such as hydrogen cyanide, oxides of nitrogen and carbon monoxide). This guideline is in operation at the Union Carbide Corporation's plant at Institute, West Virginia.

It is very difficult to understand why Union Carbide ignored the fact that MIC could degrade into Cyanide and issued a statement saying that it decomposes into "harmless substances". This statement together with the state government's confidential note banning the use of Sodium Thiosulphate gave rise to a lot of confusion and hindered the doctors from evolving a proper treatment strategy. Finally it was only seventy days after the disaster that the Indian Council of Medical Research (an apex body at the level of the Central Government that conducts research in medical sciences, referred to hereafter as ICMR), following its study conducted on gas victims admitted to an especially opened thirty bed hospital near the site of the disaster, detected that the patients treated with Sodium Thiosulphate showed remarkable symptomatic improvement and there was a clear indication of detoxification in their system. It took more than two months for the State Health authorities to accept the possibility of Cyanide poisoning. During this long impasse, the victims received all sorts of treatment except the one that would have been the most effective. As a result hundreds of people that might have been saved perished and many more suffered severe health problems.

Mystery surrounded the studies that were being conducted on the impact of the gas on various parts of the body such as the liver, kidney, brain and the blood manufacturing organs. The ICMR told the doctors who were investigating the health effects of MIC poisoning not to disclose any of their findings to the public. The Director General of ICMR announced that there was no possibility of any long-term harmful effects of MIC poisoning even before the commencement of detailed investigations. In the second fortnight of January, the Dean of the medical college of Hamidia Hospital told a gathering of doctors representing the private medical practitioners of Bhopal, not to disclose any facts relating to MIC poisoning to anyone except the State Government. Due to this, neither the clinical findings nor the X-ray pictures could be made public and the flow of medical information virtually stopped.

In these circumstances deaths continued unabated and each day the number of the dead mounted. Week after week, new patients were coming to the hospitals. They complained of pain in the chest, coughing and watering of the eyes. Those discharged earlier began returning for treatment. The doctors soon noticed that many of those who had inhaled a very small amount of the gas did not feel its effects immediately but with the passage of time it had started showing side-effects.

Victims who were not fatally affected have either gone blind or are suffering from pulmonary and other serious disorders. I came across patients with cerebral edema who had been in a coma for more than a month. Cerebral damage was also found in those victims who died within three to five hours of exposure to the gas.

In the first few hours after the gas leak women complained of abdominal pain and those pregnant among them had enough cause for anxiety. The percentage of still births and abortions among women rose sharply after the leak and this trend continued till much after. Many children "who were the most affected" were in coma for a long time and some of them never recovered. I also met many troubled parents who told me that their children had started getting up from their beds screaming in the middle of the night.

Moving out of the hospitals, I saw the other aspects of the tragedy. It was a common sight to see grim and saddened faces peering at posters showing the photographs of the unclaimed bodies searching desperately for their friends and loved ones. Crowds of people could be seen running in despair from one hospital to another and then to the police stations looking for the names of their relatives in the official list of those who had died on that fateful night. Walking along the affected areas, I sadly saw the trail of misery that the disaster had left behind.

I met Munawwar (a resident of Jai Prakash Colony, house no.16), a seventy year old man who had lost his son Ansar, a youth in his twenties, under tragic circumstances. He related that he along with his son fell exhausted on the roadside as they fled from the poisonous gas. While they lay on the road, four people picked them up and carried them to the Hamidia Hospital. He recalled how he was treated in the overcrowded passage, and his son was taken inside the medical ward. This was all he could remember as he fell unconscious after that. When he regained his senses and enquired about his son, he found to his utter dismay that nobody could tell him where his son was. Ever since, the old man has been running from pillar to post looking for his son. He was at a total loss to understand that if his son had succumbed to the gas, why was his dead body not handed over to him?

Umesh Singh, another resident of Jai Prakash colony told me that he had lost his father. Few houses away, I met Satish, a five year old boy, the only survivor in a family of eight. Further down the road in Kenchi Cholla I came across Kamala who till the gas leak lived with her three daughters while her husband Kishanlal was under detention. The gas claimed the lives of two of her daughters and Kamala is now left with her youngest child. In the same locality, Rukmini Bai is the lone survivor of a family of four. Her father and two of her brothers perished on that deadly night. The disaster has left behind endless sorrow.

I was also told that many of the inhabitants of the affected localities were found missing. Residents to whom I talked felt that many of those who had collapsed and died while fleeing from the affected areas during the early hours of the morning were lifted and their bodies disposed off at State initiative and no trace was left of them. Many young girls who had got lost in the stampede while running for their lives that morning were not traceable even months after the disaster.

The people of Bhopal have gone through endless suffering. The State relief program has throughout remained subject to bitter criticism. Relatives of most of those who died out of the hospitals and in other towns or those who could not procure certificates from government hospitals have been denied relief.

Parvati's story (resident of house no.244, Jai Prakash Colony) is a sad reflection of the indifferent and inefficient way in which the State Government has implemented its relief programs. She had a close-knit family. There was Hemraj - her husband, 8 year old son Mukesh and her daughter Usha, barely a year old. On the fateful night the couple suddenly woke up with a feeling of extreme suffocation and the unexpected noise of people running on the streets. In no time they were out on the road with their children in search of fresh air and soon found themselves running with the crowd for safety. After they had gone a few hundred yards, they dropped on the pavement as they could run no longer. Their eyes were burning and each moment they were finding it hard to breath. Before sunrise, they found that Mukesh's condition had worsened, he just refused to respond to any stimulus, there was froth all around his mouth and his face was covered with blisters. The couple took the help of the passers-by to rush him to the Hamidia hospital. The effect of the gas was so severe that Mukesh collapsed before the doctors attended to him. The grief stricken couple left with their daughter for their village Hinauti, 20 km from Bhopal, for the last rites of their dead son. But Parvati's sorrows did not end there. There was more grief in store for her. Her husband also died the next day due to the poisonous effects of MIC. Fifteen days later she returned to her forsaken dwelling, holding her baby close to her bosom. Parvati's suffering is immense but she has nothing to look forward to. The much talked about relief has evaded her because she does not have any official record to justify her claim for financial help.

A TOTAL LIE

The phase immediately following the disaster was replete with conflicting opinions about the true identity of the gas and its harmful effects. Some of the chemicals involved in the process employed by Union Carbide at its Bhopal plant were carbon monoxide, chlorine, phosgene, MIC and alpha nephthol. Since all these are highly toxic substances, there was confusion about the gas that had escaped into the air. The problem was magnified many folds because Union Carbide had always maintained a high amount of secrecy about its manufacturing process. Matters became further complicated as the company kept issuing contradictory statements. Every newspaper carried diverse reports by scientists. Most of the toxicologists in India were of the opinion that the gas that killed thousands of people was in all probability Phosgene - which until then was considered more deadly than MIC. Medical experts were also confounded - N.P. Mishra, a senior cardiologist at the local Gandhi Medical College said that most of the children and older people had died due to lung failure caused by inhaling MIC but at the same time he did not rule out the possibility of the presence of Phosgene; his assessment was based on autopsy reports. While controversy was raging, the Vice Chairman of Union Carbide India Limited told a press conference that at the time of the leak there was zero stock of Phosgene in the plant.

The involvement of phosgene could not be discounted - given the process used for manufacturing Sevin: " Phosgene combines with a chemical called methylamine to make MIC from which is produced the pesticide Carbaryl known by the trade name of SEVIN." As a matter of fact, scientists and experts who conducted studies about the effect of gas on human body, plants and vegetation were convinced that a large amount of MIC and a very small quantity of phosgene could have caused the tragedy. As far as phosgene is concerned, it is well known to be an extremely toxic substance. The Germans used it with devastating effect at Ypres in 1915 and since then it has been listed as a chemical weapon. On the contrary, the toxic severity of MIC was largely unknown till the time of the Bhopal disaster. Unfortunately, thousands of people had to pay with their lives before the scientists were to realize that MIC is five times more toxic than Phosgene.

The confusion was not just about the identity of the gas that had leaked that night. The state government's posture following the leak was also astonishing. The authorities withheld vital information about the effects of the gas on human beings, birds, animals and vegetables. All investigative findings were kept away from doctors and scientists and no guidelines for treatment were available. The wind direction data was also not made public. The government repeatedly assured the people that air, water and vegetation were safe for consumption. Even Prime Minister Rajiv Gandhi made a similar statement during his trip to Bhopal on 4th December (the day after the poisonous gas leak). The assurance from Rajiv Gandhi was rather hurried and premature and did not become his office. His statement came even before the experts who were conducting the tests on air and vegetation had reported their findings. As a matter of fact, samples of air and water for testing the effect of the gas were not collected until 5th December and it was only two days later that Indian agricultural scientists left for Bhopal to examine the crops standing in the fields. According to a Press Trust of India news release of 7th December, vegetable samples brought to Delhi from Bhopal were declared safe for human consumption simply on the basis of a "visual examination", this in spite of the fact that the leaves had been found to be badly damaged. The same news also added that "a detailed analysis of chemical residues would be completed only the next day". The claim that the vegetables were safe for consumption was completely undermined by reports that the high ranking officials and ministers were receiving their stock of vegetables from places far away from Bhopal. The residents of Bhopal found the flow of information, both from the authorities and the media, utterly misleading. As a result, the administration very soon lost all credibility among the people.

At the same time word went round that there was a likelihood of another leak and one had no reason to rule out a repetition of what had happened on the midnight of 2nd and 3rd December. The very fact that there was sufficient stock of MIC still left at the plant was cause for serious concern. Due to these reasons rumours floated freely and the city was gripped by fear. At this point, instead of informing the people of Bhopal that there were safe methods of exhausting the stock of the gas at the Carbide plant; the authorities busied themselves at gaining maximum political advantage out of the grave situation. Arjun Singh, the then Chief Minister, lost no time in declaring that the plant would never be allowed to operate in Bhopal thus hoping to win the confidence of the people. However, the employees of Union Carbide disclosed that the plant was continuing production, even after the gas disaster, in order to consume the remaining stock of the poisonous gas. [6]

This information spread throughout the city like wild fire. In an attempt to counteract the tense situation, the government announced that all talk of activities inside the plant was baseless. It however admitted that according to the experts, converting MIC into the end product was the best way to neutralize the remaining stock. This obviously implied that the plant would have had to start again (though for a limited period), contrary to the Chief Ministers earlier announcement that the Carbide plant was closed once for all.

Finally, on 11th December, the Chief Minister personally informed the people that scientists had been working on a "zero risk method" to neutralize the remaining MIC and they had concluded that it would have to be converted into 'Sevin Carbaryl'- the end product. He further said that there was no need to evacuate the city during the neutralization process but he added that the gas neutralization programme would not start without a seventy-two hour prior warning. The public at this stage was baffled because on the one hand there was talk of a "zero risk method" and on the other hand there was this warning that the neutralization process would not start without a substantial notice.

While the citizens were getting renewed assurances from the government about total safety, rumours to the contrary were spreading and even a vigorous effort on the part of the government to project that all was safe had no effect on the citizens. Some of the actions taken by the government contributed to this loss of trust. For instance, on 11th December, the State Government suddenly announced the closure of all schools and colleges in the city till the 23rd of the month. In another move a fleet of three hundred buses of the State Road Transport Corporation were kept ready at Bhopal. These buses arrived from all parts of the State to take part in a possible emergency evacuation. As a standby arrangement, special trains and additional railway coaches were also brought to Bhopal. Employees on official tours or on leave were summarily called back to the State capital. All this within a matter of twenty-four hours.

On 12th December, it was announced that the process to neutralize the remaining gas would commence on 16th morning. The lid was finally off and fear engulfed the town. Disregarding all assurances, people thronged the railway station and the main bus station to flee from Bhopal. Within a day, more than fifty thousand people left the town. During the next three days, the city witnessed an exodus that left it barren and desolate. Those who could not find transportation fled the city on foot. It was a common sight to see hand driven carts rolling down the highways - whole families with their meagre belongings perched on them. The railway station was flooded with people. Passengers barged into the trains even before the trains came to a halt at the station. Everyone was desperate to leave the city at any cost and as the trains left Bhopal, they carried with them more passengers than they had ever done before. People were inside the trains, on top of them, on the platforms and everywhere the eyes could see. All the gas stations in the city had long queues of cars and two-wheelers. Everyone was finding his way to rush out of the town. Shutters had come down at every shop and the banks were left with no cash to honour the cheques. Quite a few people sold their gold ornaments at throwaway prices (as low as rupees two hundred for ten grams of gold)[1] [7]to raise funds that could last them while they were on the run. Eventually, by the commencement of the gas neutralization programme, more than three hundred thousand people had left the city! The mass exodus that lasted for three days caused untold misery and hardship.

On the eve of the commencement of gas "neutralisation" the State Government announced that it had made arrangements to accommodate all those who stayed near the Carbide plant in especially set up relief camps.[8] At the same time it arranged buses to take people free of cost to any of the district towns within the state. Air Force helicopters were also kept on alert at the airport to spray water on Union Carbide plant and the adjoining roads while the activities to neutralize the gas were on. In view of these actions of the government, any talk of total safety was unpalatable. By the time the process of gas neutralization commenced on 16th December the whole administration had lost credibility. Strangely enough, Arjun Singh called this exercise as "Operation Faith".

At last, Union Carbide became the scene of a much awaited drama. The Chief Minister made it a point to be present inside the plant at the crucial moment and it was ensured that this was given much advance publicity. As a precaution, the plant was covered by a sheet of tarpaulin; its boundary was lined with a makeshift jute wall and both these were constantly kept wet in order to check any escaping gas. Fire tenders and helicopters sprayed water all over the plant and the roads around it, throughout the operation. The production of Sevin Carbaryl went on for five days; a day longer than initially planned. Everything went according to schedule except that the stock of gas was found to be five tons more than what was assessed on the basis of the plant records. Finally, the stock of deadly MIC was done away with once for all. The fact that there was no MIC left in Bhopal, lead to an all round respite and before long people started returning to their homes.

The period between the gas leak and the commencement of "Operation Faith" was marked by a strange rhetoric and even stranger actions. The "arrest" of Warren Anderson, chairman of the Union Carbide Corporation along with Keshab Mahindra and V.P.Gokhale - Chairman and Managing Director of Union Carbide of India respectively, on 7th December at Bhopal is a case in point. One wonders what purpose if any, was served by merely holding the three of them in the comfort of their own guest house for six hours before setting them free! It appears that this action was merely an attempt on the part of the State Government to establish that the law would not spare anyone. Later, the Chief Minister of the State was to remark that Anderson was arrested for his "constructive liability" and that his arrest served the "spirit of law". Chief Minister Mr. Arjun Singh is correct in holding Anderson "constructively liable" for the disaster, but what about the "constructive liability of someone who had been at the helm of state affairs since 1980 - ever since Union Carbide started manufacturing MIC at Bhopal?

Following the disaster, the government did its level best to manipulate every bit of information to suit its political interests. Neither the intensity of the havoc nor the death tally disclosed by the authorities came near the truth. The death figures projected even two months after the tragedy were solely based on the number of autopsies conducted. The official death figures did not take into account those who had died away from the hospitals. The residents of the affected areas have testified that dead bodies were hurriedly removed from places like the railway station and the bus station at the state's initiative and secretly disposed during the early hours of 3rd December. It is feared that many persons who may have been unconscious but still alive were probably buried or cremated in the mass clearing up operations. Both the State Government and the Union Carbide were to gain from this act of "deliberate twisting of statistics". One - they wouldn't have to account for a very big disaster; and two - fewer the victims lesser the compensation.

The relief operations that began soon after the disaster left much to be desired although on the surface it appeared that appropriate measures were being taken. On 3rd December the Chief Minister promptly raised the compensation issue and declared that the State will sue Union Carbide for damages. Prime Minister Rajiv Gandhi spared twenty million rupees in three instalments from his relief fund for the aid of the gas victims. On paper the whole relief plan looked very convincing but in reality the relief activities remained far from satisfactory. The authorities, to the chagrin of the people who were affected the most, neglected them and concentrated their efforts in areas that were not affected very badly. One may wonder why anyone would do so? The general elections to the Parliament of India were round the corner and it is possible that the ruling party decided to exploit the huge monetary resources that were to be distributed to the victims. It appears that the members of the ruling political group decided that the survivors were in such a bitter state of mind that they would not support them in the forthcoming elections no matter what aid was given to them. On the other hand it might be easier to placate those who had suffered the least. It is possible that this was the reason why they concentrated all relief efforts in areas that were least affected. Their plan may have paid dividends for they won the elections. It is unlikely that all this could have happened without the higher ups in the ruling party being aware of the situation.

The supporters of the ruling Congress party, especially the members of the Municipal Corporation were also found putting pressure on government doctors to issue medical certificates to undeserving persons so that they could be entitled to receive monetary help. The doctors took a strong exception to this; they refused to bend under pressure and even resorted to a day long strike in protest. The worst instance of corruption that soon became common knowledge was seen in the distribution of interim relief. The government decided that the relatives of those who had died were to be given rupees ten thousand for every dead victim, but the agents responsible for the distribution of money unashamedly handed out merely one hundred to two hundred rupees. They could succeed in doing so because the poor and illiterate people were not aware of the amount that was due to them. [9]

The very purpose of relief was ultimately defeated and the victims out of shear frustration came out on the streets in great numbers to protest against the improper relief measures. It is sad to note that corruption has come to symbolize any activity involving money and the relief operations for the gas affected victims of Bhopal have been no exception either.

THE CULPRIT

As a witness to the poisonous gas leak and its fall out, it will be inhuman on our part to turn a blind eye to the various causes and circumstances that led to the catastrophe. The world ought to know the circumstances that enabled Union Carbide to set its foot in Bhopal; who are responsible for letting the multinational corporation establish a hazardous plant within city limits; and how the safety systems in a supposedly modern plant failed?

The answers to these questions are not difficult to find. India launched a project to fight its chronic food shortage in the mid-1960s. The plan was to usher in a "Green Revolution" that required intensive use of high yielding varieties of food grains. For the survival of crops from these hybrid seeds it became essential to use pesticides and fertilizers. Union Carbide came on the scene recognizing the potential for manufacture of pesticides in India. In 1969, it applied to the Union Government for a license to manufacture pesticide. Three years later both parties signed a letter of intent and finally, after "due care and investigations" the Union Industries and Civil Supplies Ministry issued the license [10] to the company in 1975 to annually produce 5000 tons of Sevin Carbaryl - a pesticide based on MIC.

Union Carbide India Limited's proposal for a license was processed according to the provisions of the Foreign Exchange Regulation Act (FERA) as fifty one percent of the total share capital of the company is owned by the Union Carbide Corporation. The project was screened thoroughly at many levels, both administrative and scientific. Agencies like the Indian Council for Agricultural Research and the Indian Council for Medical Research were involved at this stage to ensure that the company met the minimum safety requirements. Finally, a committee of secretaries to the government of India recommended the proposal to a select committee of the Union Cabinet for its approval. The role of the Government of Madhya Pradesh (the State whose capital is Bhopal) had also been very crucial; it allotted the site for the construction of the plant and periodically inspected the plant before certifying it safe for production through the office of the Inspector General of Industrial safety. The Pollution Control Board and the Ministry of Environment were also responsible for ensuring that the Union Carbide plant was not an environmental hazard.

Looking at the government's role we find that there has been a serious flouting of rules and safety norms at every step. The Indian authorities did not care to obtain basic information about the chemicals that were scheduled for storage and use in the plant before issuing the license. This in spite of the fact that the Canadian authorities had earlier found a similar plant unsafe and rejected it on safety grounds. The allotment of land for building the plant within eight hundred yards of the railway station was a serious offense. Sadly enough, the safety of the citizens of Bhopal was overlooked in preference to the interests of the multinational corporation.

The Union Carbide Corporation neglected all norms of safety and delivered for its Bhopal venture, a plant design with substandard manual safety devices that could not guarantee against a sudden leak. On the other hand, the same company installed a highly sophisticated computerised safety system in its pesticide plant in Institute, West Virginia. The safety record of Union Carbide's Bhopal factory, ever since it was commissioned, remained far from satisfactory. Flaws in the plant design showed time and again. In particular, three of its workers died due to exposure to Phosgene prior to the December 1984 leak. In spite of frequent accidents and warnings about the danger to the lives of the people, no corrective action was taken either by the Company or the State Administration. Within a year of its inception, higher profits became the sole consideration for the management and the economy drive that ensued, resulted in a decline in the training standards of the plant personnel. The workers were also shuffled at random from one area of specialization to another without much regard for fitness for specific jobs. All this drastically affected the efficiency of operations. Fact remains that the Carbide plant was not ununsspect till the deadly hours of the 3rd December 1984 and people were already aware of the dangers they were facing from it. Rajkumar Keshwani, a journalist from Bhopal had informed the people through a series of articles that were published in the local dailies that the Carbide plant was a big hazard. On 17th September, 1982 he wrote an article under the title "Save, please save this city" Another news story written by him appeared on 1st October, 1982. The headline read "Bhopal sitting on the mouth of a volcano." He went on to write "if you don't understand you will perish."

The writings of Keshwani had no effect on the management or the State authorities. Nevertheless, he went ahead with his crusade and personally wrote to the Chief Minister but the latter did not even think that the issue was grave enough to be addressed. Keshwani was alarmed by the frequent leaks of Phosgene at the MIC plant. The State Government had confirmed in the State Assembly that a fitter had died earlier and Mohd Ashraf had lost his life on December 25, 1981 due to leakage of gas in the factory. On 9th February, 1982 a considerable amount of phosgene had again escaped into the air and seriously affected twenty four people working inside the plant and they had to be rushed to hospital.

When the issue of gas leak and plant safety was raised in the State Legislative Assembly in 1982 and the attention of Tara Singh Viyogi, the then Labour Minister, was drawn to the Union Carbide plant, more particularly, the dangers that the citizens were facing due to the location of Carbide plant within city limits, the minister had quipped that the plant was not a stone that could be moved away at will. He also gave the assurance that there was no danger to the citizens from the gases that were used in the plant. The reply of the minister was grossly misleading and false. It seems that there were vested interests in the political set up and the bureaucracy that took care of the objectives of Union Carbide. The reply of the labour minister is a glaring instance of the unmindful government support to the multinational company at every stage.

Ultimately, a disaster like the one that struck Bhopal has made us realize that hatchet men of industrial superpowers are operating in India with the collusion of few Indians who stoop low in exchange of a few favours. The accomplices are within us and they ought to be exposed. India has the requisite scientific knowhow that can detect and prevent dangerous elements from securing licenses. There is no reason why a deadly contraption like the Union Carbide pesticide plant should have been allowed to settle on Indian soil. It seems that the findings and the suggestions of the experts were shelved at the decision stage because of other considerations.

LEGAL WRANGLING

Following at the heels of the Bhopal disaster, there appeared signs of a legal warfare, both in India and the United States - Lawyers from the U.S. aided by their Indian counterparts lost no time and landed at Bhopal to prepare for big damage suits. Within a week of the gas leak, lawsuits claiming a total of thirty-five billion dollars in damages were filed against the Union Carbide Corporation in the U.S. courts. Looking at the legalities involved, the experts rightly concluded that the legal wrangling would continue for years.

On 9th December 1984, John P. Coale[11] a well known tort lawyer from Washington D.C. was the first to arrive at Bhopal along with Arthur Lowy lawyer, and Ted Dickenson investigator. On his arrival, Coale busied himself in finalising retainer agreements with the victims and assessing the situation in the worst affected areas including the main hospital. While Coale went to Bhopal, Melvin Belli [1] a San Francisco personal injury lawyer, filed a fifteen billion dollars damage suit (five billion dollars in compensation and ten billion dollars in punitive damages) on 7th December against Union Carbide Corporation in the Federal Court at Charleston, West Virginia. Belli described the gas leak as the greatest disaster in a foreign country that involved a United States based multinational corporation, and predicted that this accident would greatly affect the way such companies do business in the third world. In his petition he said that the Union Carbide negligently failed to install the same safety warning device in its plant in Bhopal as in a similar plant in Charleston. He also blamed the company for having failed to warn the citizens who lived near the plant about the dangers of the gas.

On 10th December Belli arrived in Delhi on his way to Bhopal. There he held discussions with K. Parasaran, the Attorney General of India and the Chief Minister of Madhya Pradesh. During these meetings he impressed upon them to file a "class suit" against Union carbide Corporation in the American courts. Legal activities soon picked up a hectic pace and the rush of American lawyers continued. A California based team of lawyers including Jay Gould, Fredrico Syre and Ralph Fertiz also reached Bhopal a day before Belli's arrival. Before proceeding for Bhopal, the Gould-Syre team had already filed a twenty billion dollar compensation suit in the East District court of New York on behalf of four Indian citizens.

Soon after converging at the site of the disaster, the American attorneys set up law offices on the footpath across the Union Carbide factory. It was a common sight to watch thousands of victims as well as relatives of the deceased line up at these law offices to sign up retainer agreements for multimillion dollar law suits over the disaster. In the process, well over fifty thousand people signed up as plaintiffs in less than month's time.

While the lawyers were busy building their case for the victims, the State Government started making arrangements for filing a fifteen billion dollar suit of damages in a United States court. At the same time, the Municipal Corporation of Bhopal also signed an agreement with John Coale for filing a multimillion dollar suit in America on its behalf. The then city Mayor D.K. Bisarya said that lawyers would be seeking compensation from the giant corporation for the "huge losses° suffered by the Civic authority of Bhopal during and after the tragedy.

A week after the gas leak, the Supreme Court of India also entered the legal arena. N.A. Krishnamoorthy, an Indian attorney moved a petition in the highest Indian court to seek compensation from the government of India and the government of Madhya Pradesh, of half a million rupees for the families of each of those killed or injured. He also contended that a sum of rupees five thousand should be given to the owners of cattle that had perished due to the poisonous gas.

Krishnamoorthy's charge was that the Government of India and the government of Madhya Pradesh acted "arbitrarily" and in complete violation of the Fundamental Rights guaranteed under Articles 14, 19(1) (G) and 21 of the Constitution of India by permitting the Union Carbide to establish its plant in a heavily populated area. He said that the fundamental rights (right to life and liberty) of the victims were violated due to the negligence of the Government of Madhya Pradesh and the Union of India.

Krishnamoorhty, in his case, had levelled a definite charge that the fundamental rights of the citizens were violated and it was the responsibility of the Supreme Court to go into the depth of the case and bring the culprits to book. However, in the very first hearing the Chief Justice said "... while the suits were in the process of being filed the petition in question appeared premature in nature." Elaborating his remark further he went on to add it is a casual petition thrown at the court. No law points have been raised or cited."

A close look at the proceedings that took place on 11th December in the Supreme Court, suggests that the judges had already formed their opinion on the writ petition. While agreeing to the remarks of the Chief Justice, justice Sen commented that the State had already taken steps for providing succour to the victims (his claim was far removed from reality). At the same time, Justice Eradi's remark that the petition was "bare of facts" and his subsequent query, How is the State Government guilty of negligence?" left little doubt in one's mind that a decision favourable to the victims was unlikely.

After barely three hearings, Krishnamoorthy decided to withdraw his case,[12] even though the Supreme court had served notices on the Union of India, Union Carbide India Limited and the State of Madhya Pradesh, besides granting him the permission to move an application to make the multinational Corporation a party in the petition.

As the days passed hectic activities continued on the legal front. Lawyers from the U.S. signed up thousands of plaintiffs and filed damage suits in the United States against the Union Carbide Corporation. During a long period which lasted over hundred days, the Government of Madhya Pradesh made concerted efforts to file a separate damage suit in the U.S. courts against the multinational company. The Attorney General of India was sent to the U.S. to discuss various legal issues with the American attorneys. In the meantime, the President of India in a surprise move, promulgated an ordinance enabling the Union of India to acquire the "rights of a guardian" to represent the victims of the gas disaster in the U.S. courts, to secure compensation from the Union Carbide, or to enter into an out of court settlement on behalf of the victims. The Indian Parliament subsequently passed the Bhopal Gas Leak Disaster (processing of Claims) Bill replacing the Presidential ordinance.

After the Act [1] [13]was passed, the Indian government filed a law suit[14] against the Union Carbide Corporation, holding it "absolutely liable" for the gas disaster. The suit filed in the United States' Southern District Court of New York, demanded unspecified sums to compensate the victims and punishment to the defendants, so as to deter the multinationals from "wanton disregard of the rights and safety of the citizens" The Indian government also asked the court to give them the control of the jumble of law suits that had been filed. Ignoring the Parliamentary Act, the gas victims went ahead and approached the American attorneys individually to file cases on their behalf. The U.S. courts admitted the applications of the Indian citizens ignoring the latest Act that was passed by the Indian Parliament, as they (the courts) were not bound by the Indian Act. At the same time, the victims also filed law suits in Indian courts, against the Union Carbide India Limited, the Union of India and the State of Madhya Pradesh holding them guilty for the disaster.

The major legal questions that arose out of all this activity were - whether the damage claims should be heard in the United States of America or should they be heard in India, which country's law should govern the legal proceedings, whether the claims should be against the parent company or the Indian subsidiary etc. As far as the legalities were concerned, parent company, its Indian subsidiary, as well as the licensing authority in India should have been held liable under different counts.

Under American and British jurisprudence, the strict and absolute liability under the rules laid down in Rylands vs Fletcher have been expanded to what may be described as "all dangerous operations", including the "escape of dangerous gas or fire". According to the Rylands case, it can be laid down categorically that a poisonous gas is a dangerous thing and hence a person who keeps or brings such a gas on to his premises shall be governed by it. The conclusion is obvious that he who keeps the gas keeps it at his own peril, and furthermore, even the authority granting the permission to store such gas would be liable if the gas causes any harm.

According to the legal provisions of the government of India Air Act, 1981 and the Insecticides Act of 1968, the representatives of the government, that is, the ministers and the officials who were responsible for ensuring compliance with the legal provisions of the Acts, can be held liable and are punishable for failing to ensure that the Carbide plant operated under proper safety conditions. There was a clear case of criminal negligence on the part of both the Central and the State governments. Since the offence was under the special legislation like the Pollution Act, the principle of strict liability could have been applied without proof of mens rea

In this context it is appropriate to consider an important ruling of the Chief Justice of the Supreme Court of India on 1st August 1983 in the famous case of Rudal Shah, who had been illegally imprisoned for over fourteen years. In his ruling, Justice Chandrachud said: "Article 21, which guarantees the right to life and liberty, will be denuded of its significant content if the powers of the Supreme Court were limited to passing orders of release from illegal detention. One of the telling ways in which the violation of that right can reasonably be prevented and due compliance with the mandate of article 21 secured, is to mulct its violators in the payment of monetary compensation. Therefore, the State must repair the damage done by its officers to the petitioner's rights. It may have recourse against those officers".

The concern of the Chief Justice, towards guaranteeing respect for the citizen's right to life and liberty, is distinctly clear from his ruling. His verdict helps in concluding that the Plaintiff has a right to sue the state, its ministers and officers or both.

In the case of the Bhopal disaster, it will be unfortunate to overlook the utter negligence on the part of the Union Carbide Corporation, Union Carbide India Limited, the Union of India and the government of Madhya Pradesh. There is no doubt that each of them is liable for compensation and punitive damages. The personnel who were involved in the decision making process and are guilty of negligence should be criminally prosecuted.

With definite charges against itself, including "contributory negligence", the government of India made unnecessary attempts to file a suit against the multinational corporation, when the right to sue was solely the victims' right.

Following the tragedy, efforts were made to confuse the citizens of Bhopal about their legal rights. The people were misled by the Parliamentary Act that gave to the government of India all authority to represent their case in the U.S. given that the government was itself liable for legal action. As a consequence of concerted media efforts, the attention of the survivors finally centred on the compensation issue. The victims concluded that if they had to receive any compensation it was to come from the Union Carbide Corporation. As far as Union Carbide is concerned it made every effort to escape criminal charges levelled against it. The multinational corporation repeatedly expressed its desire for a mutual settlement. Its keenness for a compromise was explicit from the statement made by its Chairman soon after his return trip from Bhopal which was cut short by his arrest. Anderson had asserted at a hurriedly called press conference at Danbury on 10th December that an "equitable resolution of the compensation question was possible without resort to the path of litigation which was time consuming and would not benefit the victims of the tragedy."

John Coale (who was retained by more than six thousand victims to fight their case in the U.S. courts), held the same view. Upon his arrival in Washington D.C. from Bhopal on 12th December, he said: "we had rather not go through a long drawn out battle in the courts; we had rather get the relief quickly. That's why we have not filed suit" He stressed, in particular that he represented the vast majority of those with damage and that his eventual settlement offer would probably be a much smaller figure. Making his stand categorically clear, he stated, "not in the billions, I don't think so".

The similarity in the sentiments expressed by Anderson and Coale were cause enough for concern. It raised doubts about the attorney's integrity towards those whom he professed to represent. As a matter of fact, the Indian government's initiative to be the sole representative of the victims in the U.S. courts also suited Union Carbide's interests as it did not have to convince thousands of victims individually to accept a compromise.

As far as compensation is concerned, both Union Carbide and the respective governments in India (state as well as the Union), agree that relief and succour should be provided to the victims and their relatives, but the crucial issue is: Does justice end at compensation? Shouldn't the victims be awarded punitive damages besides compensation?

On the legal front, a year after the disaster, writ petitions were pending in the Indian and the U.S. courts against the Union Carbide and the Indian Government. The hearings in the U.S. courts began with much speculation about the final verdict. However, the outcome of the very first hearing held on 16th April 1985 in the Federal Court of New York, was encouraging for the survivors of the gas leak. The Judge John Keenan directed that prior to the final outcome of the case; emergency relief should be disbursed to the victims under a special relief programme. (Shouldn't the Supreme Court of India have taken a similar stand in response to the writ petition filed by NA. Krishnamoorthy). Subsequently, as late as the first week of January 1986, the legal proceedings still remained where they had begun. It was undecided whether the cases were to be heard in the U.S. or in India.

JUSTICE DELAYED

In March 1986 a news item appeared in an American newspaper disclosing that Union Carbide Corporation had agreed to pay dollars 350 million as compensation to the gas victims. The story went on to add that the Indian Government was likely to accept the out-of-court settlement offer. Later, when Union Carbide realised that it would be difficult to negotiate with the Indian government, it mutually worked out a compromise settlement offer of dollars 350 million with a 3-members committee of US attorneys representing a large number of victims in the US Courts. The committee included Stanley Chasley, Lee Beily and Michael Circessi, Chasley's group, in particular, was representing over 50,000 gas victims. The compromise offer was nothing but an attempt by the multinational corporation to put pressure upon the government to reach an out-of-court settlement. In this design, the Union Carbide succeeded to the extent that it could muster the support of some of the Attorneys representing the gas victims. Later, Stanley Chasley's collegues, John P Coale and Arthur Lowy reached Bhopal in the last week of March to discuss the compromise offer with the victims. At a press conference called by these attorneys in Bhopal on 29th March, 1986, when I asked Coale why he had failed to demand punitive damage from the multinational company before accepting a proposed settlement offer, Coale's reply was that punitive damages were not possible since deliberate negligence on the part of Union Carbide Corporation was difficult to prove. He also added that he had reached this conclusion on the basis of "thorough investigation". Besides, he aired his full throated contention that an out-of-court settlement for dollars 350 million was a huge amount and it should be accepted. There was widespread criticism of this offer and the

Government of Madhya Pradesh even issued an appeal to the Bhopal gas victims not to be misled by certain unscrupulous elements who were inducing the gas victims to accept a nominal compensation instead of what they justly deserved. The State Government's appeal described as misleading the reported comments of the American attorneys that if the settlement was not reached at dollars 350 million, judge Keenan would send back the compensation case to India. Judge Keenan was critical of a letter written by Coale and Lowy to Madhya Pradesh Chief Minister Motilal Vora in which the attorneys had said that they were guaranteeing dollars 350 million to the Bhopal victims. Judge Keenan called this letter "an outrage" and said that the letter was misleading and confusing. The judge said that there was only a proposal to settle the case and that nothing had been settled finally by the Court. Indian government's outright rejection of the proposed settlement offer was understandable since the amount of dollars 350 million fell short of what was fair and equitable. While the situation became increasingly rife with speculations about the final outcome of the case, Judge Keenan declared on 12th May, 1986 that India and not the US was the proper forum for such litigation and this verdict no doubt was in accordance with the multinational's preference for a trial in India. The decision of Judge Keenan was a setback to the Indian Government as well as the victims. The reassuring feature of the judgement was that the judge had imposed firm conditions on the Union Carbide Corporation. The verdict passed by the Federal Court of New York required the multinational corporation to submit to the Indian Court's jurisdiction waiving limitations of statutes, and subject itself to the process of discovery of documents and other evidence under the US rules and procedures, as well as to the judgement of Indian courts. Later, Union Carbide Corporation, left with no alternative agreed on 12th June, 1986 to accept the conditions laid down by the Federal judge of New York district court in order to have law suits over the Bhopal Chemical disaster moved from the US Courts to India.

[1] Arjun Singh went to the extent of referring this matter to the (Justice N.K. Singh) enquiry Commission.

[2] At the time of the gas leak more than 30,000 gallons of MIC was stored at the Union Carbide Plant at Bhopal, whereas, the storage capacity is not more than 5000 gallons at the Carbide Plant in Institute, West Virginia and that too with much more sophisticated safety mechanisms.

[3] On 3rd evening I was attended to at the Jai Prakash Hospital and was advised to take complete rest. Later after investigations I was told by the doctors that my upper respiratory tract was affected and my eyes were superficially damaged by exposure to MIC.

1. [4] It was found that each of the above was also accompanied by superficial keratitis in which only the exposed portions, i.e., inter palpebral fissure were affected.

[5] Doctors noticed patches of pneumonitis. There were increased markings going up to the lateral one-third of the chest and lung fields, besides there were punted minute opacities with shadows all over the lung fields. In some cases there was also patchy consolidation resembling broncho pneumonia and pulmonary edema in the most severe ones. The pulmonary edema was bilateral and maximal. There was crown glass haziness and pair - bronchial cuffing and these readings were sufficient in the process of diagnosing pulmonary edema.

[6] This fact was later verified upon investigations. The records at the Electric sub-station that supplied power to the Carbide plant

confirmed that the plant consumed the same amount of energy during the period in question as it did while its MIC plant was in operation prior to the gas leak.

[7] The market price was rupees two thousand for the same amount of gold.

[8] All was not well with the camps and many unfortunate women and young girls who stayed there went through a never ending trauma for they were physically assaulted in the camps. Even certain policemen, posted there to ensure their safety, were involved in these inhuman acts.

[9] This became a widespread practice and the matter was also brought to the notice of the state government. When the officials were asked why the amount given to the victims was written only in figures (which could easily be altered) and not in words while issuing the receipts, their reply was that since there was a heavy rush of sufferers they did not have the time to follow all the formalities! In this way the enquiry was concluded at the highest level of bureaucracy and the erring officials went scot free

[10] The license no. being C/11/401(75) dated November 31st, 1975 - signed by N.K. Berwa, Under Secretary to the ministry.

[11] John P. Coale represented the Americans who were held hostage in Iran from 1979 till 1981.

[12] Krishnamoorthy's writ petition No. 17268/84, NA Krishnamoorthy Vs State of Madhya Pradesh and others was permitted to be withdrawn, by the order dated 4th April, 1985, of the Chief Justice, V.V.

[13] This Act was challenged in the Supreme Court of India by a number of citizens as violative of the Fundamental Right guaranteed by the Indian Constitution.

[14] The Indian Government was represented in the US by Robbins, Zelle, Larson and Keplan of Minnesota.

www.ingramcontent.com/pod-product-compliance
Lightning Source LLC
Chambersburg PA
CBHW070335240526
45466CB00027B/1986